Distance

Marie Lemke

A Crabtree Roots Plus Book

Crabtree Publishing
crabtreebooks.com

School-to-Home Support for Caregivers and Teachers

This book helps children grow by letting them practice reading. Here are a few guiding questions to help the reader with building his or her comprehension skills. Possible answers appear here in red.

Before Reading:
- What do I think this book is about?
  - *I think this book is about how far things are from each other.*
  - *I think this book is about measuring distance.*
- What do I want to learn about this topic?
  - *I want to learn how to measure distance.*
  - *I want to learn about miles and kilometers.*

During Reading:
- I wonder why…
  - *I wonder why a car uses an odometer.*
  - *I wonder why walking one kilometer is faster than walking one mile.*
- What have I learned so far?
  - *I have learned that the United States uses inches, feet, yards, and miles.*
  - *I have learned that I can use yardsticks and metersticks to measure short distances.*

After Reading:
- What details did I learn about this topic?
  - *I have learned that monarch butterflies fly a long distance.*
  - *I have learned that the North Pole is a long distance from the South Pole.*
- Read the book again and look for the vocabulary words.
  - *I see the word **ruler** on page 10 and the word **odometer** on page 12. The other vocabulary words are found on page 23.*

I can measure distance.

Distance is the space between two things.

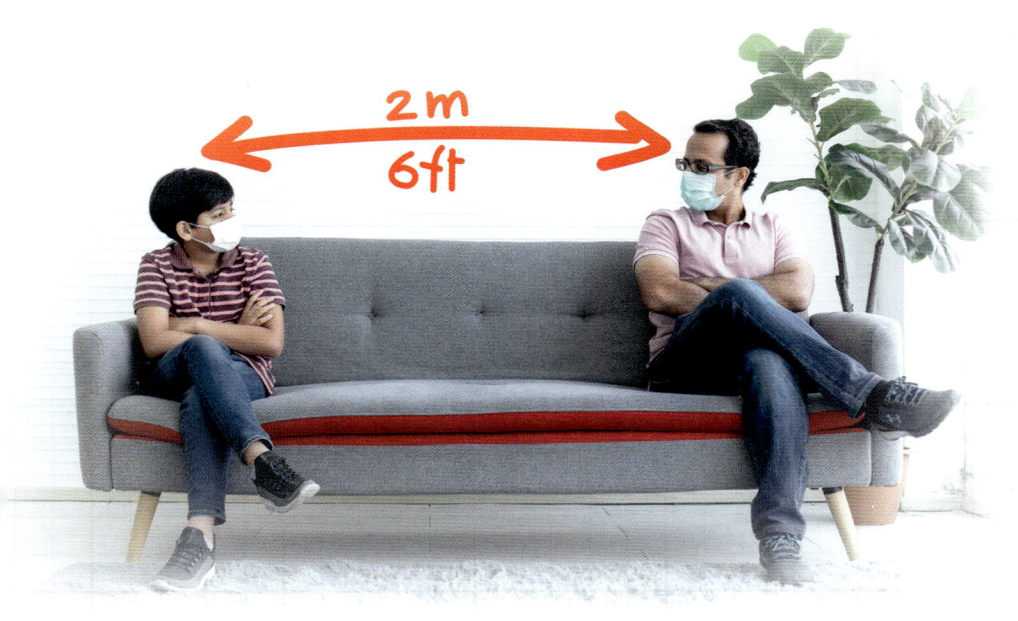

It tells me how far apart two things are.

There are different ways to measure distance. The imperial system uses inches, feet, yards, and miles. The United States uses this system.

1 foot = 12 inches

1 yard = 36 inches

1 yard = 3 feet

1 mile = 5,280 feet

1 mile = 1,760 yards

Most of the world uses the metric system.
It uses meters and kilometers.

10 millimeters = 1 centimeter

100 centimeters = 1 meter

1 meter = 1,000 millimeters

1,000 meters = 1 kilometer

A **ruler** measures short distances. Yardsticks and metersticks are for short distances, too.

Cars use **odometers**. They show the distance driven in miles or kilometers.

I can walk one mile in 20 minutes.

I can walk one kilometer in 10 minutes.

Chicago is about 700 miles (1,126 km) from New York City.

The **moon** is 238,855 miles (384,400 km) from Earth.

The **North Pole** is 12,430 miles (20,004 km) from the South Pole.

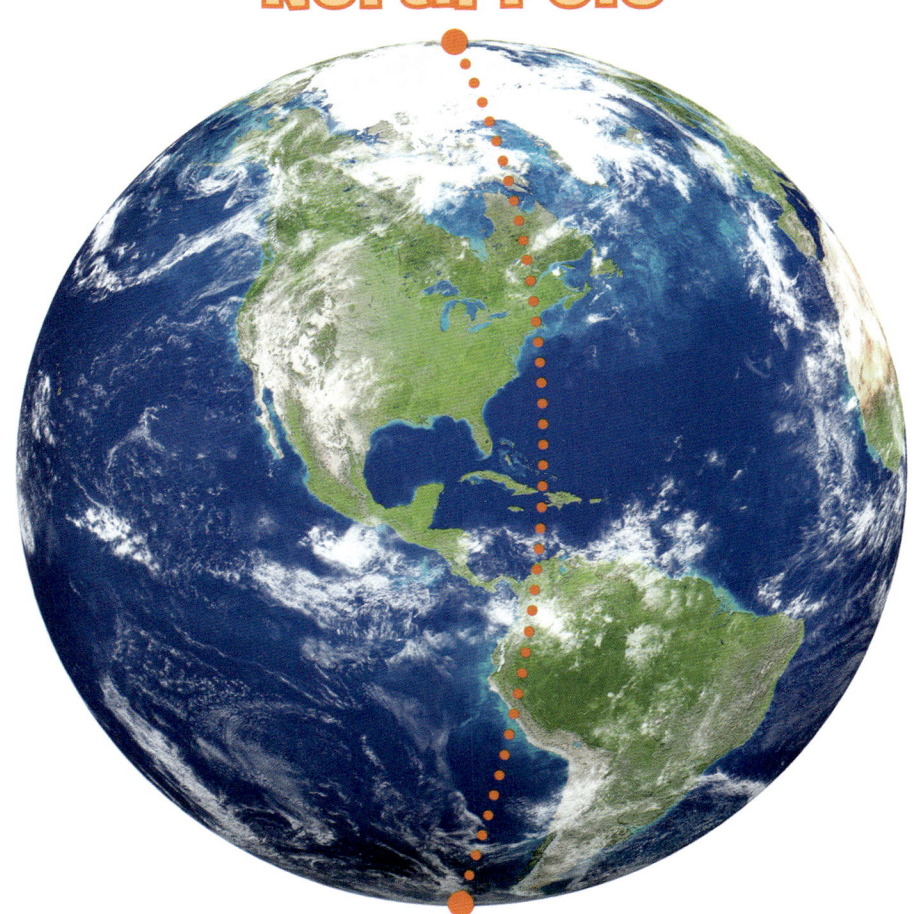

Monarch butterflies fly from Canada to Mexico every year. They travel about 3,000 miles (4,828 km).

Word List
Sight Words

a	far	most	to
about	fly	of	too
and	for	one	two
are	from	short	very
between	how	show	walk
can	I	tells	world
car	is	the	use
city	it	they	year
Earth	me	things	
every	mile	this	

Words to Know

Chicago

monarch butterflies

moon

North Pole

odometers

ruler

Measuring Things
Distance

Written by: Marie Lemke
Designed by: Jen Bowers
Series Development: James Earley
Proofreader: Janine Deschenes
Educational Consultant: Marie Lemke M.Ed.

Photographs:
Shutterstock: Cover: tape measure ©2010 Bragin Alexey, image ©2018 Ruslan Shugushev; p.2&3 binder clips ©2021 Irina Gutyryak; p.3 ©2018 Dmitryp-k; p.4 ©2015 Maria Symchych; p.5 ©2021 chomplearn; p.7 paper note ©schab; p.10 ©2017 Green Leaf; p.11 ©2021 Alex from the Rock; p.13 ©2016 Belle Ciezak; p.14 ©2013 Viorel Sima; p.15 ©2020 ASDF_MEDIA; p.16 ©2017 bobby20; p.17 ©2009 marilyn barbone; p.19 ©2008 Alex Staroseltsev; p.21 ©2018 Isabelle OHara

Crabtree Publishing

crabtreebooks.com   800.387.7650

Copyright © 2023 Crabtree Publishing

All rights reserved. No part of this publication may be reproduced, stored in a retrieval system or be transmitted in any form or by any means, electronic, mechanical, photocopying, recording, or otherwise, without the prior written permission of Crabtree Publishing. In Canada: We acknowledge the financial support of the Government of Canada through the Canada Book Fund for our publishing activities.

Printed in the U.S.A./012023/CG20220815

Published in Canada
Crabtree Publishing
616 Welland Ave
St. Catharines, Ontario
L2M 5V6

Published in the United States
Crabtree Publishing
347 Fifth Ave
Suite 1402-145
New York, NY 10016

Library and Archives Canada Cataloguing in Publication
Available at the Library and Archives Canada

Library of Congress Cataloging-in-Publication Data
Available at the Library of Congress

Hardcover: 978-1-0396-9645-7
Paperback: 978-1-0396-9752-2
Ebook (pdf): 978-1-0396-9966-3
Epub: 978-1-0396-9859-8